CARPENTRY & CABINET-MAKING

REAL LIFE GUIDES

Practical guides for practical people

About this series

In this increasingly sophisticated world the need for manually skilled people to build our homes, cut our hair, fix our boilers, and make our cars go is greater than ever. As things progress, so the level of training and competence required of our skilled manual workers increases.

In this new series of career guides from Trotman, we look in detail at what it takes to train for, get into, and be successful at a wide spectrum of practical careers. *Real Life Guides* aim to inform and inspire young people and adults alike by providing comprehensive yet hard-hitting and often blunt information about what it takes to succeed in these careers.

The other titles in the series are:

Real Life Guide to Catering

Real Life Guide to Construction

Real Life Guide to Hairdressing

Real Life Guide to Plumbing

Real Life Guide to the Motor Industry

trotman

Real Life GUIDES

CARPENTRY & CABINET-MAKING

Dee Pilgrim

Real Life Guide to Carpentry and Cabinet-Making
This first edition published in 2003 by Trotman and Company Ltd
2 The Green, Richmond, Surrey TW9 1PL

Editorial and Publishing Team
Author Dee Pilgrim
Series Editor Timuchin Dindjer
Editorial Mina Patria, Editorial Director; Rachel Lockhart, Commissioning Editor; Anya Wilson, Editor; Erin Milliken, Editorial Assistant.
Production Ken Ruskin, Head of Pre-Press and Production
Sales and Marketing Deborah Jones, Head of Sales & Marketing
Managing Director Toby Trotman

Designed by XAB

British Library Cataloguing in Publication Data
A catalogue record for this book is available from the British Library

ISBN 0 85660 909 9

All rights reserved. No part of this publication may be reproduced, stored in a retrieval system or transmitted in any form or by any means, electronic and mechanical, photocopying, recording or otherwise without prior permission of Trotman and Company Ltd.

Typeset by Photoprint, Torquay
Printed and bound in Great Britain by The Cromwell Press, Trowbridge, Wiltshire.

Real Life GUIDES

CONTENTS

ACKNOWLEDGEMENTS	6
ABOUT THE AUTHOR	6
INTRODUCTION	7
1. SUCCESS STORY	11
2. WHAT'S THE STORY?	15
3. TOOLS OF THE TRADE	19
4. A DAY IN THE LIFE OF A JOINER	27
5. MAKING YOUR MIND UP	31
6. CASE STUDY 1	35
7. TRAINING DAY	39
8. CASE STUDY 2	47
9. CAREER OPPORTUNITIES	51
10. THE LAST WORD	55
11. RESOURCES	57

Acknowledgements

Thank you to Viscount Linley, Marc Smith, Darren Minett, and Paul Magerum for their contributions to this book.

Many thanks to the Carpenters' Company and staff at the Building Crafts College.

For detailed information about training in the industry, a big thank you to the Construction Industry Training Board (CITB), and most especially to John Carter.

About the author

Dee Pilgrim completed the pre-entry, periodical journalism course at the London College of Printing before working on a variety of music and women's titles. As a freelancer and full-time member of staff she has written numerous articles and interviews for *Company*, *Cosmopolitan*, *New Woman*, *Woman's Journal* and *Weight Watcher's* magazines. As a freelancer for Independent Magazines she concentrated on celebrity interviews and film, theatre and restaurant reviews for such titles as *Ms London*, *Girl About Town*, *LAM* and *Nine To Five* magazines, and in her capacity as a critic she has appeared on both radio and television. She is currently the Film Reviewer for *Now* Magazine. When not attending film screenings she is active within the Critics' Circle, co-writes songs and is currently engaged in writing the narrative to an as yet unpublished trilogy of children's illustrated books.

Introduction

It's a cry to be heard from homeowners the length and breadth of the country these days: 'Why can't I find a good carpenter when I need one!' Why, indeed, when those with woodworking skills are now assured of constant employment anywhere in the United Kingdom. You may think of a carpenter as a Bob the Builder character, travelling around in his van fitting new windows, cupboards and doors in people's private homes, but this is only one of a myriad of different jobs carpenters undertake. The truth is there are shortages of skilled tradesmen in all areas. In fact,

People who decide to train to be a carpenter can now get highly paid for their skills.

according to a report by the Construction Industry Training Board (CITB), 56,000 more carpenters and joiners are needed to enter the industry by 2006 – that's over 18,500 new carpenters each year.

The problem really started some time ago. John Carter, the National Training Advisor dealing with shopfitting at CITB, explains how different the industry used to be. 'It was very boom and bust', he says, 'and so there was a fluctuating workload due to government policy and a number of other factors. Because of this fewer people were being recruited and educated in

Conservation is one of the biggest growth areas. A lot of money has been pumped into conservation by English Heritage, Heritage Scotland and by the Church and they need really skilled workers.

woodworking skills because they wanted a more secure job and working future. Also, a lot of skilled workers retired and were not replaced as firms became leaner and fitter in order to stay in profit. It's also true to say that the cultural outlook has changed. Where once craftsmen were admired for the skills they had and the job they did, youngsters who could have come into the industry began to think it was too much like hard work what with the getting up early, working out in all weathers, and still having to go to college to train when they could be sitting at a desk in an office pressing buttons.' Another problem was that, once trained, carpenters would get poached by other industries that could offer better wages and conditions. 'In the past we would lose our new trainees to other jobs where they used their manual skills – like the car industry in Birmingham – that paid about three times what carpentry could. That has changed because we now have better pay and conditions, but we are still losing a lot of trained carpenters from the construction site to boat building, especially canal boats and other small pleasure craft. All these factors have led to a shortage across the board in woodworking and the shortage has created a situation where people who decide to train to be a carpenter can now get highly paid for their skills.'

This shortage has been highlighted by the huge amount of construction that is now happening, which is mainly because of changes in our living arrangements. Since the Second World War there has been a huge change in the way people live in this country. At one time the norm was for young people to live with their parents in the family home until they got married. Now, young people tend to move away from home much earlier. Many young people want to live independently, either in shared private accommodation or in houses or flats they have bought or rent alone. People are also getting married much older and are more likely to buy property by themselves before they marry. The population is also growing steadily, which has led to a huge demand for new properties to be built around the country. In the

South East alone, the government has set aside three-quarters of a billion pounds for building affordable housing over the next three years. That figure will rise to more than £20 billion over the next 15 years. That's an awful lot of houses for carpenters to work on.

There has also been an upturn in the construction of commercial property, such as office blocks (anyone who has visited Canary Wharf in London's Docklands will have been overawed by the amount of construction there), and large public amenities such as hospitals. According to John Carter, another area that has expanded considerably is the renovation and upkeep of older, or 'heritage', buildings. 'Conservation is one of the biggest growth areas,' he says. 'A lot of money has been pumped into conservation by English Heritage, Heritage Scotland and by the Church and they need really skilled workers.' There are currently nearly a quarter of a million carpenters and joiners working in Britain. It is estimated that by the end of 2003 over 1.48 million people will be employed in the construction industry, with a fair proportion of them involved in some way in working with wood.

Another area where there has been great expansion is jobs for local freelance carpenters. The massive rise in interest in DIY and house decoration has actually done carpenters a favour and has certainly raised the carpenter's profile – look at the celebrity status achieved by Handy Andy from the television show *Changing Rooms*. Although there are plenty of DIY jobs the amateur feels he or she can tackle, there are many more they can't, and this is where the skilled carpenter comes in. New wood floors, a beautifully turned balustrade for the staircase, or refitting traditional sash windows are just some of the projects for which you need a well-trained carpenter. There is also a growing demand for female carpenters. 'We are aware there are not enough females working in the industry and we have started to target women returners [to work after an absence],' says John

Carter. 'More women in the industry has many benefits, not least because a lot of them do freelance work and as so many women now live on their own in their own houses they feel more secure hiring a female carpenter to work for them.'

The interest in interior design has also re-sparked an interest in the skills of the cabinet-maker, with increasing demand for beautiful, individual, handmade pieces of furniture. It's not surprising that this increase in demand for woodworkers has come at the same time as a renewed public interest in wood. Ten years ago plastic, metals and other durable materials were all the rage, and no-one would have dreamed of putting timber decking in the garden, while wood panelling on walls was seen as very old-fashioned. But now wood, in all its glorious colours and textures, is making a comeback into our homes. All of this means it has never been a better time to train in carpentry, for although the work is highly skilled and demanding, the rewards – both personal and financial – can be immense.

> **DID YOU KNOW?**
>
> Carpenters who make wooden guitars by hand are known as luthiers.

This book is intended to help you decide whether a career in carpentry is for you. It will explain what the various branches of carpentry actually entail and what skills you will need to enable you to do the job. It will offer information on what qualifications you will need to progress to a vocational course in carpentry and also what the broader opportunities in carpentry are. Whether you want to work as one of a team, or for yourself and by yourself, working with wood to produce something practical, such as a roof beam, or something beautiful, such as an exquisite marquetry table, can be a wonderfully fulfilling way of earning your living. And remember, there's somebody out there right now crying out for a good carpenter!

VISCOUNT LINLEY

Success story

THE CABINET-MAKER

David, Viscount Linley, is the Queen's nephew and currently one of the foremost furniture designers working in wood in England. After discovering an interest in designing and building desks and tables as a schoolboy, he went on to win a place at Parnham House, the school for craftsmen in wood. Here he learned about all aspects of his trade and after leaving in 1982 he started to work at a co-operative in the village of Betchworth. His workload quickly increased with numerous commissions for bespoke pieces of furniture. In 1984 he launched his first shop, David Linley Furniture, in London's New King's Road, and during the next few years completed projects as diverse as the Linley broadwood piano, a boardroom table for the Metropolitan Museum in New York, and a marquetry spiral staircase for Bowater House. The shop moved to larger premises in London's Pimlico Road in 1993 and David's first book, *Classical Furniture*, was published in the same year. In 1996, his second book, *Extraordinary Furniture*, was published and in 1997 he collaborated with the Sèvres porcelain factory to produce a cabinet, Sèvres' first piece of furniture since the nineteenth century. In 1998 the shop was re-launched and David set up the Linley-Rycotewood Scholarship at

> Go and work in a factory or in a workshop, just try out different things and you'll discover if it is for you.

Rycotewood College. Since then he has published another book, *Design and Detail in the Home*, and launched various new collections of furniture including upholstered chairs and sofas. He now employs 44 staff in London and subcontracts a lot of work to craftsmen around the country, most especially at his workshop in Whitby.

> If I was to start all over again I'd certainly either try to be an apprentice or start in a workshop. I would have liked to do that because you build up a really deep knowledge of how a workshop actually operates, of how your particular bit works.

'I enjoyed woodwork, design, pottery and art at school because there was always that satisfaction at the end of the day of being able to take an end product home. This was particularly important to me because both my parents were visual and they enjoyed seeing what I had made.

'From school I went straight into cabinet-making and I missed out on carpentry really and that was a shame because carpentry would have been a very useful supplement to have learned. I went straight into the cabinet-making side of things where you learn to make boxes and how to make dovetails, but carpentry would have been a very good stepping stone to being a cabinet-maker because carpenters know all sorts of ingenious tricks! If I was to start all over again I'd certainly either try to be an apprentice or start in a workshop. I would have liked to do that because you build up a really deep knowledge of how a workshop actually operates, of how your particular bit works.

'At Parnham we spent the whole of the first year understanding the practical side of the business and we also got taught secretarial skills and business skills but there really wasn't enough of that. When we left, after two years, we were perfectly capable of making things but we really didn't know how to balance our budgets or how to pay the bills and I think those are good skills to have.

'Cabinet-making is the most complicated you can get in terms of woodwork and it is certainly challenging. You can spend hours trying to work out how you are going to glue a certain piece up. At the beginning I was working with immensely talented people and came to realise you have to concentrate on the things you are good at, so I decided to focus on producing designs and marketing them and subcontracted the actual making of the designs to a network of craftsmen. There is an innate skill in employing people and I feel immensely proud about the relationships we have built up. We have worked alongside many of our craftsmen pretty much since we started and helped and developed and given advice where they have asked for it, and it is very much a partnership in terms of understanding what we are both trying to achieve.

> Get as much experience as you can before you start. Be as prepared as you can and look into every avenue of what to expect.

'We have to face the possibility that less and less people will have their skills and abilities. In the last 20 years we have probably seen the greatest change in this industry there has ever been. Up until then carpentry and cabinet-making were done the traditional way but now everything is computerised, everything is bar-coded and everything is automated, so a lot of the skills have been removed in order to employ unskilled people who cost less, which makes the company more profitable, but makes no sense in the long term. I think people are beginning to appreciate what

DID YOU KNOW?

Celebrity actor Harrison Ford actually worked as a carpenter in Los Angeles before hitting the big time. His carpentry skills stood him in good stead when he was working on the film "Witness", where in one scene he helped erect a timber frame house.

we do because so few people actually know how to do it any more. That's why I've set up the scholarship – to help people come out into the world slightly more educated, prepared and experienced.

'The best advice I can give is get as much experience as you can before you start. Be as prepared as you can and look into every avenue of what to expect. Go and work in a factory or in a workshop, just try out different things, and you'll discover if it is for you. I discovered I didn't like working in a factory environment by actually going and doing it during the school holidays. Also, there are huge amounts of books on design and decorative arts out there that will help you be aware of what is current. You need to look at the market and know what is selling. You also have to be adaptable, you have to rise to the challenge and must never give up. Finally, be ambitious. I'm ambitious and I don't think you can do this if you are not. If you don't want to do things then it just won't happen.'

secretarial skills and business skills but there really wasn't enough of that. When we left, after two years, we were perfectly capable of making things but we really didn't know how to balance our budgets or how to pay the bills and I think those are good skills to have.

'Cabinet-making is the most complicated you can get in terms of woodwork and it is certainly challenging. You can spend hours trying to work out how you are going to glue a certain piece up. At the beginning I was working with immensely talented people and came to realise you have to concentrate on the things you are good at, so I decided to focus on producing designs and marketing them and subcontracted the actual making of the designs to a network of craftsmen. There is an innate skill in employing people and I feel immensely proud about the relationships we have built up. We have worked alongside many of our craftsmen pretty much since we started and helped and developed and given advice where they have asked for it, and it is very much a partnership in terms of understanding what we are both trying to achieve.

> Get as much experience as you can before you start. Be as prepared as you can and look into every avenue of what to expect.

'We have to face the possibility that less and less people will have their skills and abilities. In the last 20 years we have probably seen the greatest change in this industry there has ever been. Up until then carpentry and cabinet-making were done the traditional way but now everything is computerised, everything is bar-coded and everything is automated, so a lot of the skills have been removed in order to employ unskilled people who cost less, which makes the company more profitable, but makes no sense in the long term. I think people are beginning to appreciate what

DID YOU KNOW?

Celebrity actor Harrison Ford actually worked as a carpenter in Los Angeles before hitting the big time. His carpentry skills stood him in good stead when he was working on the film "Witness", where in one scene he helped erect a timber frame house.

we do because so few people actually know how to do it any more. That's why I've set up the scholarship – to help people come out into the world slightly more educated, prepared and experienced.

'The best advice I can give is get as much experience as you can before you start. Be as prepared as you can and look into every avenue of what to expect. Go and work in a factory or in a workshop, just try out different things, and you'll discover if it is for you. I discovered I didn't like working in a factory environment by actually going and doing it during the school holidays. Also, there are huge amounts of books on design and decorative arts out there that will help you be aware of what is current. You need to look at the market and know what is selling. You also have to be adaptable, you have to rise to the challenge and must never give up. Finally, be ambitious. I'm ambitious and I don't think you can do this if you are not. If you don't want to do things then it just won't happen.'

What's the Story?

2

If your parents have ever had new cupboards built at home, or your school has had wooden window frames replaced, you will have seen a carpenter at work first hand. However, the maintenance worker or freelancer who travels around building or fixing wooden structures such as kitchen cabinets, staircases and box sash windows is only the tip of the iceberg: there's much more to the carpentry story than that. Carpenters are everywhere, even when you are not aware of their activities, and they need to be highly trained and skilled in order to do their jobs. But what does being a carpenter or a joiner mean? What about a cabinet-maker? Basically, carpentry is the art of cutting, joining and framing together timbers essential to the stability of a structure, while joinery is the art of dressing, framing, joining and fixing wood for the finishings of houses. So, in basic carpentry, it doesn't really matter what it looks like as long as it does its job, while joinery not only has to do its job, it also has to look good because it will be on view. A cabinet-maker either puts together ready-cut parts of furniture in a workshop, or is a skilled joiner producing hand-made desks, chests of drawers, and other pieces of furniture of high quality.

Nearly every building you see around you, even those that are predominantly constructed from concrete, metal and glass, will have required the services of a carpenter at some point while being built. Here are the most common places you will find carpenters, joiners and cabinet-makers at work, with a brief description of what their different titles and specialities are.

CONSTRUCTION SITES

This is where **formworkers** make temporary structures that are used like moulds to support wet concrete before it sets, **joiners** erect and fix floor joists and roof timbers, **first fix carpenters** finish off the internal woodwork, such as partition walls, and **second fix carpenters** do the final bits, such as skirting boards, door surrounds and fitted wardrobes.

WORKSHOPS

The workshop is where **bench joiners** construct fitted furniture, panelled doors, windows, and staircases, and where the **cabinet-maker** makes anything from desks to chests of drawers.

MACHINE SHOPS

Machinists prepare and shape rough timbers into floorboards, skirting boards, and panels using specialist equipment.

TV/FILM/THEATRE SETS

When scenery has to be built for a TV production or a film it may be constructed in sections in a specialist workshop before being erected on a stage or in a studio. Alternatively, it may be built in the studio itself.

ON SITE

Freelance carpenters and **shopfitters** (who specialise in producing the fronts and interiors for restaurants, banks, and shops) may do most of their work on site.

Obviously, carpenters who want to do the more intricate, skilled jobs need a lot more training and experience than those whose duties are quite basic. However, there is no clear route of progression through carpentry: this will very much depend on your own skills and interests. For example, you may decide to become a **construction manager**, responsible for keeping a job on budget and meeting deadlines while overseeing a construction crew, and for this you will need a university degree or many years of experience in the business. Alternatively, if you like the security of working regular hours for someone else, you may decide to become a **maintenance carpenter** working for a local authority, for example, where your duties will involve the upkeep of public buildings such as schools or council offices. For this career, vocational qualifications are more appropriate and there are a number of ways you can train.

The traditional way to train as a carpenter has always been to become an apprentice, working for a company or individual while learning on the job. This has benefits for the company, in that it gets an extra pair of hands to help out with the more menial tasks, and for the apprentice, who receives a wage, or at least expenses, while learning. Learning while working is still the favoured method of training for carpenters and joiners. The Training Day chapter (page 39) looks at the training and qualifications available in more detail.

> **DID YOU KNOW?**
>
> Between 2001 and 2002 the number of manual trade trainees (including carpentry trainees) doing the CITB's Construction Apprenticeship Scheme went up 10 per cent from 8,294 to 9,106.

3

Tools of the trade

When people think of the tools used in carpentry they usually visualise a workroom stocked with saws, drills, hammers and chisels. Of course, these are all vital in woodworking and a carpenter could not get by without them. Yet there are other, more personal, tools that are just as important. While hard work and determination are needed in order to progress in any job, there are more specific skills and strengths that a carpenter will rely on throughout his or her working life. By pinpointing these skills you will be able to see whether you really do have what it takes to make a career in carpentry. If you believe carpentry is for you, you can purchase hammers and saws as you train and work, but for now, see how many of these personal tools you already possess.

- A carpenter does not spend their working hours sitting behind a desk – they are more likely to be on their feet making it! This will involve picking out timbers, bringing them to the workshop, sawing them, joining them and finishing them. All this hard physical work can be pretty demanding, so being physically strong and having stamina are essential for a job in carpentry. **Strength and physical fitness** are even more important for carpenters who work on building sites, erecting roof timbers or wooden frameworks for houses.
- If you decide to become a carpenter working on the construction of buildings, you really do need to have a **head for heights**. You could well find yourself on the top of a two-storey building fitting roof timbers, or up a ladder fixing the wooden panelling on a gable end.

- Although you will be taught the practical skills of carpentry during your training, it does help if you are already quite deft. Good **hand-eye co-ordination and dexterity** will certainly make your life easier, especially if you decide to specialise in the more intricate sides of carpentry such as finishing, cabinet-making or marquetry.
- It's not just your hands that have to be deft: you'll be using your brain quite a lot to work out how much wood you need for a certain job, what angle a joint should be and the correct thickness of certain timbers and panels. This means you have to be good at mathematics and geometry, and extremely precise with your measurements. You also need to be able to read and correctly interpret construction pictures and diagrams. It's no good building a lovely wooden door only to find it is too big to fit in the doorframe. **Accuracy** is essential.
- Anywhere that construction is taking place and power tools are being used can be dangerous. A good carpenter is always conscious of the **health and safety** aspects of each individual job. In carpentry hazards lurk everywhere. Building timbers are heavy, you'll often be working high up on ladders or scaffolding, circular saws are very sharp, and electric power tools must be treated with the utmost respect. This is particularly true on building sites where heavy plant is also in use, so being constantly aware, cautious and careful are great skills to have.
- Being able to understand what other people are talking about and making yourself understood in return are essential to carpenters. If you are working as part of a team you need to know who is doing what and when, especially as you will be liaising with other skilled workers such as plumbers and electricians. (Pipes and wires will have to be installed behind that lovely wooden wall panelling.) Even freelance carpenters who work alone need to have good **communication skills** as they will be ordering raw materials from suppliers and talking to their clients about their needs and requirements.

Instructions need to be clear and concise so that everyone knows exactly what is going on.
- Having **tact and diplomacy** will make your job much easier. If a delivery fails to arrive, or a customer is unhappy with a job it is no use screaming and shouting about it. You have to co-operate with other people and if they can see you dealing with difficulties politely and efficiently they are much more likely to use you again themselves, and to recommend you to other people.
- **Punctuality** is vital wherever you are working. If deliveries are being made to the workshop, you have to be there to make sure they are actually what you ordered. If you tell a client you will be arriving on site at 9am sharp, they will not be amused if you turn up at midday. People lead busy lives and haven't got the time to sit around waiting for you to arrive when you feel like it. Being punctual is a sign of professionalism, so use it to your advantage.

We've looked at the skills you need in order to get on: now let's look at certain physical conditions that could hold you back, and aspects of working in carpentry that may not suit you personally. Think about these seriously before making up your mind that this is the career for you.

- If you suffer from **vertigo**, a building site will most certainly not be for you: carpenters are required to climb up scaffolding and ladders in order to fit timbers on roofs. Just think of the carpenters who have to fix church spires!
- Working with wood creates a lot of dust, and even though safety standards and measures have improved dramatically in recent years, if you suffer from **breathing problems** or have allergies to dust this could cause difficulties. No matter how good your face mask or the ventilation in the workshop, it is inevitable that some wood dust particles will remain in the air.
- Many different adhesives are used in carpentry, for gluing joints, and for fixing wood veneer and laminates. If you suffer

from **skin sensitivities** these can cause allergic reactions such as contact dermatitis.
- Many carpenters find they are forced to change job due to **back problems**. Although the correct and least damaging ways to lift and carry wood are taught in college, anyone with a family history of back trouble should be aware that bad backs are common in the industry.
- If you can't add up to save your life or think geometry is an alien language, you won't be able to cope with the complicated equations necessary to produce a curved wooden structure such as an arched doorway or garden bridge. Mathematics is part of the job, so **if you hate maths** this really isn't for you.
- Many carpenters still work in the traditional way, 'following' the work around the country. This means they **travel** all over Great Britain and abroad, working on various building projects as they arise. If you don't want to spend long periods of time away from home, this particular type of carpentry work will not suit you.

You should now have a better idea of some of the skills and strengths needed for the job and some of the downsides to the industry. If you are still enthusiastic about carpentry, complete the short **quiz** below to discover what your general knowledge of the work is like. All you have to do is choose the multiple-choice answer you believe is correct in each case. This is really just a fun way of seeing if you know as much about carpentry as you think you do, so don't worry if you get some wrong. Answers appear at the end of the quiz.

1. Which of the following tools would you NOT expect to find in a carpenter's workshop?

    A. Hammer
    B. Chisel
    C. Rotary whisk.

2. You are building a bespoke kitchen for a client and he gives you a sample of the wood he wants you to use to make his work surfaces. You approach your usual supplier who has the right quantities of that particular wood in stock, but it is a totally different colour from the sample. Do you:

    A. Go ahead and order the timber anyway? Well, it's what the client said they wanted.
    B. Go back to the client and make sure this is the timber they want you to use?
    C. Contact other timber merchants in the area to see if they have the correct coloured wood in stock?

3. What is the correct amount of wastage you should estimate into any job?

    A. None, you are sure of your measurements
    B. 10 per cent
    C. 50 per cent.

4. You have produced an estimate for a client on quite a big and costly job because you are using expensive oak timber. However, after you've done the estimate and before you can purchase the timber the price of oak suddenly rockets. Do you:

    A. Immediately contact the client and explain what has happened?
    B. Go ahead with the job?
    C. Get on the Internet and see if anyone has stocks of oak at lower prices?

5. Which of the following is not a type of hinge?

    A. Piano
    B. Tenon
    C. Rising butt.

6. You are about to use a power tool on site. It runs through a transformer, but what is the correct voltage?

   A. 50 volts
   B. 110 volts
   C. 420 volts.

7. Which of the following is not standard safety gear for a carpenter?

   A. Steel-toe-capped shoes
   B. Facemask
   C. Ear defenders.

8. You have a job in a private house where the client is a keen DIYer. He keeps telling you he would not do things the way you are doing them and seems to want you to carry out the job his way. Do you:

   A. Explain diplomatically that you are a trained professional and you know exactly what you are doing?
   B. Do it his way just to shut him up?
   C. Agree to do it his way but say that will add extra costs to the job and so you will have to increase the bill?

9. Which of the following statements does a carpenter live by?

   A. Measure once, cut once
   B. Measure twice, cut once
   C. Don't bother to measure, estimate.

10. You have a particularly difficult client who keeps changing the spec of the job. Do you:

    A. Lose your temper and walk off the job?
    B. Just make the changes she wants because you can sort it all out at the end?
    C. Sit the client down and discuss the job until you and the client know exactly what it is she wants?

ANSWERS

1. C. You would find a rotary whisk in a kitchen, not in a carpenter's workshop.

2. B. You really need to talk to the client before you go any further. Wood, being a natural material, does vary in colour, but if the wood your timber merchant has in stock is the wrong shade it may not suit the overall design of your client's kitchen. Then again, they may prefer this colour and ask you to go ahead with it. If not, you are going to have to find a supplier who does stock this particular shade of wood, or get your client to choose an alternative.

3. B. You should always factor in a degree of wastage. This is because there may be imperfections in the wood and the shape of whatever you are making may produce offcuts that are of no use to you. Ten per cent is the industry standard.

> **DID YOU KNOW?**
>
> Marquetry, the word for inlaid wood, comes from the French word marqueter, meaning variegate.

4. A. This is another occasion when you really need to talk to the client immediately because they may not have the budget to afford the more expensive oak. If they wish to proceed with oak, then you can start looking for other, cheaper suppliers, otherwise you'll have to look for an alternative timber.

5. B. A tenon is a joint, not a hinge.

6. B. There is a reason why power tools work off 110 volts. If you were to get an electric shock of 110 volts it would be rather uncomfortable, but it would not kill you. However, 420 volts most certainly would!

7. This is a trick question: all the answers are correct.

8. A. Always remember, you are a professional who has trained to do this job. Explain to the client that you are doing things the way you are because this is the correct method and it will produce the best results. Never compromise your standards.

9. B. 'Measure twice, cut once' is like a mantra to a carpenter. Accuracy is paramount. You do not want to waste your time and resources by having to discard wood because you have cut it to the wrong size and have to start again.

10. C. Once again, it is time to discuss things with the client. There is no point starting a job only for the client to say it is not what she wants and forcing you to start all over again. You have to know exactly what it is she has in mind before you begin. If she is still not sure you need to be polite but firm and insist that you need precise instructions.

Having completed this section of the book you should have a better idea if you have the skills and strengths to be a carpenter or cabinet-maker. If you are still certain about wanting to enter the industry the next chapter will explain what could be in it for you personally.

4

A day in the life of a joiner

Darren Minett, who is 21 years old, has been a freelance joiner for two years. Before that he was apprenticed to Barratt Homes in the Midlands for four years and studied up to NVQ3. He now works for a small joinery firm that has its own workshop and its own building arm.

'My day starts at 8am when I drive myself in my van to the workshop. When I get there the firm tells me which job I will be on that day – it could be that I will be making something in the workshop or that I will be out working on site. Today I and the other joiner in the firm have been working on a fitted kitchen for a private client. Yesterday we fitted all the units and wall units and today we've gone in and mitred the worktops and fitted all the cornicing and things like the light rail. We also fitted all the drawers and all of the doors. This really was a two-man job because some of the units were too big to handle alone.

'We are meant to have our first break at half past nine, with lunch at half twelve, but we

> Every day is different. I could never work in a factory. Going to the same place every day would really bore me.

don't always stop to have breaks if there is something we really need to get done. There is also another break at quarter past three, before we knock off at five, but if we have to work through we get paid overtime or can have time off in lieu.

'If I'm building something in the workshop we have all of our own machinery for cutting it and planing it down such as band saws and cross cut. I have made everything from doors, airing cupboards and gates to flights of stairs. When I was an apprentice Barratt's bought me a small bag of tools and I bought a few cordless as I trained. Then, when I went self-employed I went out and bought all my power tools and a bigger van so I can transport everything I need from the workshop to the site.

'If I'm out on site then the first thing I have to do is go and have a look. I have to do all the measuring up and work out what materials I will need, then I go back to the workshop and see if we have the timber in stock, work out how long the job is going to take us, and then do a cost estimate. If the client agrees to that then we book a date in for when the job can be done.

'If I am working on a building site the job I like to do the best is roofing. Heights don't bother me and I like it because it is outside and you can really see the job appear. First there's nothing, then there it is, a roof. Not many joiners can do proper, traditional roofing these days and it is hard physical work. The biggest roof job I have ever done was putting in a set of oak roof joists. The timbers were twelve-by-

> I have to do all the measuring up and work out what materials I will need, then I go back to the workshop and see if we have the timber in stock ...

fours and each timber was five metres long. It took five of us to lift each one, they were so heavy!

'At the moment I am spending every Wednesday at college doing an Advanced Vocational Certificate of Education (AVCE). This is a kind of gap two-year course filling in between doing my NVQs and going on to do an HNC in business studies. This is because in the long run I would like to have my own business. The course is really hard because it covers things like town planning and building regulations as well as all the other aspects of the business.

'The other days of the week, I might nip back to the workshop once I've finished just to sort out materials for the next day's job. That's what I love about what I do, there is so much variety and so much freedom. Every day is different. I could never work in a factory. Going to the same place every day would really bore me.'

5

Making your mind up

By now you will be aware that there's much more to carpentry than there first appears to be. Working with your hands and actually creating things can give you a powerful sense of achievement, but what else will a career in carpentry bring to your life? What are the benefits, financially, socially and personally? This section consists of a number of the most commonly asked questions about getting a job in the industry and will help you to make up your mind if this is really the career path for you.

ONCE QUALIFIED, CAN I MOVE UP THE PROMOTION LADDER QUITE QUICKLY?

That depends on what you decide to do in carpentry. Once your basic training is finished, you will have to decide what area of carpentry you want to specialise in. You could join a bespoke cabinet-makers or a shopfitting firm. You could decide to go into the construction side, working on site. Alternatively, you could decide to go it alone. Traditionally, after their apprenticeship has finished, newly trained carpenters are encouraged by their firms to leave and seek employment with someone else in order to gain more experience in a different work environment. This means you can move up the promotion ladder quickly as you master new skills. However, this is a trade in which you never stop learning, so a quick promotion may not always be best. Sometimes it is better to stay in a job and learn absolutely all you can before moving on.

WILL I WORK NINE TO FIVE?

Again, that will depend on which branch of carpentry you decide to pursue. Although some carpenters and cabinet-makers, especially those working in factories, do work a normal 39-hour week from Monday to Friday, many do not. For example, carpenters working on building sites will want to make use of all the hours of daylight they can. This means in the summer they will start earlier and finish later than normal and may work weekends as well. Many carpenters are working to a deadline and might have to work overtime to get the job finished. The freelance carpenter working in people's homes may have to tailor their hours to suit their clients.

WILL I GET TIME OFF FOR HOLIDAYS?

Yes, you will. Once again, the amount of holiday you get will depend on which branch of carpentry you are in, on your employer, and on whether you are self-employed or not. If you are self-employed you will have to calculate how much time you can take off, not only in financial terms, but also in terms of how many job opportunities you are going to miss. If you are employed on site it is normal to dovetail holidays between completing one job and starting the next.

HOW MUCH CAN I EXPECT TO EARN?

Pay will depend on your experience and expertise. Traditionally, apprentices are not particularly well paid, but this situation is improving. At the time of writing trainees at the Building Crafts College are being paid between £100 and £200 per week, depending on the employer. When fully trained, the basic rate with local authorities is about £11,200, while the craft rate is about £13,580. A more experienced carpenter can expect to earn over £19,500, while experienced cabinet-makers can earn over £16,000. However, these are only basic wages. Many woodworkers increase their earnings by working overtime or by claiming travel allowances. If they are working in London they may also be paid London Weighting.

WILL I HAVE ANY COSTS WHILE I TRAIN OR WHEN I AM FULLY QUALIFIED?
Yes, you will. Although colleges and most firms will supply you with safety gear such as goggles or face masks while you train, you will probably have to buy your own steel-toed safety boots. However, these are a great investment as you'll most certainly need them once you are qualified. You may also have to contribute to the cost of any expensive hardwoods you use at college – woods such as oak or beech are very costly compared to pine and if you are making a hardwood cabinet or shelving unit as part of your coursework the college will probably supply the wood at cost price. Obviously, you will use a lot of tools while at college and it is a good idea to start buying your own tools for use once you are qualified. You will need saws, chisels, hammers, screwdrivers and drills. If you eventually decide to become self-employed you will need either a work van or, at the least, an estate car in which to travel with your materials to worksites. You will also have to invest in power tools such as sanders.

WILL I BE ABLE TO USE MY SKILLS ABROAD?
Yes. Great Britain is not the only country with a shortage of skilled craftspeople, and many carpenters go abroad to work on contracts. The television series *Auf Wiedersehen Pet* followed the adventures of a number of British construction workers travelling to Germany to work on a site there. If you are contemplating using your skills abroad, you may like to think about taking a GCSE in a foreign language.

WHAT CAN I EXPECT TO GET OUT OF THE INDUSTRY PERSONALLY?
A great sense of satisfaction and achievement as you can actually see the finished product of your labours. Talk to any carpenter and they will tell you the most satisfying thing about their job is the fact that at the end of the day they can actually see what they have created. Working with your hands can give you a powerful sense of achievement. Also, if you work on site or in a workshop,

there is the camaraderie that grows when you and your workmates are all concentrating on a common goal. For the self-employed freelancer there is the opportunity to work in a variety of different locations and to meet a cross-section of the public.

Finally, these are skills you will never lose, so you'll never have to employ a carpenter to hang doors, fit windows or put down decking in your own home – you'll be able to do it all yourself.

> **DID YOU KNOW?**
>
> Although men usually enter training to become carpenters straight from school, most women are in their early to middle twenties before they start their training.

HOW WILL THE WIDER PUBLIC SEE ME?

As a valuable member of society, offering a professional skill. Most people are unaware of the vast variety of jobs carpenters undertake. They usually only become aware of what they do when they need one to come and work on their properties! Every homeowner who has ever needed new cupboards built, or sash windows fitted, knows the value of a well-trained, professional carpenter, as does anyone who has employed a cabinet-maker to build a special item of furniture. As so many people no longer have woodworking skills, those who do become a precious commodity. No wonder the modern carpenter is so very much in demand.

COULD I BE MY OWN BOSS EVENTUALLY?

Of course you could. Many carpenters run their own businesses. This is especially true of freelance carpenters and those who are sub-contracted to do work for a larger firm. Even if the thought of running your own business, with all the paperwork that entails, doesn't appeal to you, there are great opportunities to progress from a craftsperson up to the level of site manager or construction manager as there is currently a shortage of people with the necessary expertise.

MARC SMITH
Case study 1

THE STUDENT-TURNED-TEACHER
At just 19 years old, Marc Smith is already Assistant Bench Joinery Instructor at the Building Crafts College in Stratford, East London. He always knew he wanted to do something practical as a career, although he did not do Woodwork at school, opting instead to do Home Economics at Leigh City Technical College in Dartford, Kent. He then went on to do a GNVQ in Travel and Tourism before deciding to become a carpenter.

He set up an apprenticeship with a local shopfitting firm in Dartford before leaving college and then, at 16, he started to work for the firm and to study for his NVQ 2 and NVQ 3 in Wood Occupations at Building Crafts College. During this time, he did a block release course, working in the industry for six weeks, going to college for two weeks, then returning to work for another six weeks. He received a wage from his firm, financial assistance with fares to college, and a tool allowance.

After finishing his apprenticeship in 2002 he left his original company and started working for another shopfitting company, before returning to his original firm where he became the assistant to the Project Manager, concentrating on estimating,

6

> Communication skills are vital if you want to be a carpenter and having good maths skills really does help, but being interested is the biggest thing.

quoting and surveying. He joined the teaching staff at the Building Crafts College in March 2003.

'My dad's in the building trade and I always knew I wanted to do something practical too, which is why I got the apprenticeship. Building Crafts College was a great place to learn because the facilities are good and so is the teaching. My group consisted of people with all levels of ability: people who were academic and those who weren't, those who were good practically and others who were not so good. It was a close-knit group, even though we only saw each other for a few weeks each term, but we worked well together and everyone mucked in and helped each other.

> **DID YOU KNOW?**
>
> The Worshipful Company of Carpenters is one of the City of London's oldest livery companies, dating back to 1271.

'We were learning a wide range of skills, but my instructor told me he thought I would always be a better theory carpenter than practical carpenter and I suppose I always looked to go into the industry at a higher level, such as a site manager or site foreman, rather than as a basic carpenter. I believe some people are natural carpenters and I'm not a natural, but I was prepared to work hard and I did exceptionally well, so when the opportunity to come back to the college to teach came up it was too good an opportunity for me to miss.

'Teaching is more the theory side. You don't have to physically make a door from start to finish. What you do have to do is be able to talk to people, to explain to other people how to do it. You have to show them where they are going wrong and how they can put it right. It's one thing to be able to do it yourself, but it is another skill entirely to explain to someone else how to do it.

'In fact communication skills are vital if you want to be a carpenter and having good maths skills really does help, but being interested is the biggest thing. If you are interested and realise you will always keep learning then there is a big future for you in this industry and it's a great way to make a living. I would say to anyone, it is worth giving it a go. There's a great sense of teamwork and a wide variety in what you actually do. But best of all is if you make something, a cabinet say, your work can actually be seen. You can see your finished product and so can other people.

> I always looked to go into the industry at a higher level, such as a site manager or site foreman, rather than as a basic carpenter.

'I would definitely like to progress with the teaching and I'm hoping later to apply for my assessor's qualification and do some teaching units. I'm really enjoying it and I get great job satisfaction out of teaching. I think the future of the industry is the young people who want to come into it. If guys come away from here after three years and say "yeah, I've learned a lot" that's a real kick to me because I have a lot of time for guys who want to learn something and do something with their lives. There's a lot of job satisfaction in seeing people progress.'

Training day

By now, you should have a good idea if a job working with wood is for you. If you have decided it is, then you will need professional qualifications in order to advance as quickly and as far as you possibly can. Although there is no age limit for entry into the industry, most people start straight from school and this is because some training schemes do have an upper age limit. Carpentry has traditionally been a craft learned through an apprenticeship, in which a carpentry company or older, trained individual takes you on and teaches you the skills of the trade as you work. This is still the main way to get in to the industry and you can become an apprentice by approaching a company direct and becoming apprenticed to them or by joining a training programme such as the Construction Apprenticeship Scheme (CAS) run by CITB (see below). You can also attend a college as a full-time student, or do a degree course at university. All these methods of training lead to qualifications recognised by the industry.

NATIONAL VOCATIONAL QUALIFICATIONS

National Vocational Qualifications (NVQs) or Scottish Vocational Qualifications (SVQs) allow you to learn practical skills on the job while training at college or a training centre, usually on a block-release or day-release basis. A two-year training programme will usually lead to NVQ/SVQ 2 with the option of completing a third year up to level NVQ 3.

MODERN APPRENTICESHIP

A Modern Apprenticeship (MA, or Skillseekers in Scotland) is an extremely good way of learning your trade – so good that around 30,000 young people in the UK start some form of apprenticeship each year. It lets you earn as you learn and get qualifications. A Foundation MA will lead to NVQ/SVQ Level 2, while an Advanced MA will lead to NVQ/SVQ Level 3. You are eligible to do an MA if you are between 16 and 24 years old.

CONSTRUCTION APPRENTICESHIP SCHEME
In England and Wales a Construction Apprenticeship Scheme (CAS), run by the CITB, lasts for three years, while in Scotland there is a four-year apprenticeship registered with the Scottish Building Apprenticeship and Training Council. For both, entrants are registered with an employer and have an Apprenticeship Agreement that guarantees employment while they work towards NVQ/SVQ Levels 2 or 3. You must be between 16 and 25 to be eligible for a CAS. The CAS complements both the Foundation MA and Advanced MA in England, as well as the MA schemes in Wales.

CITY AND GUILDS
Many of the City and Guilds certificates, although not part of any national qualifications framework, are equivalent to NVQ/SVQ. You can take City and Guilds certificates in Basic Carpentry and Joinery Skills, and Carpentry and Joinery. City and Guilds also offers a two-year, full-time Diploma course. City and Guilds 'own brand' qualifications come in five levels, each level being roughly equivalent to the same level of NVQ or SVQ.

SCOTTISH QUALIFICATIONS AGENCY
For those wishing to become cabinet-makers, the Scottish Qualifications Authority (SQA) offers a Higher National Certificate (HNC) and a Higher National Diploma (HND) in Furniture Restoration and Furniture Design.

BTEC
BTEC (now run by Edexcel) offers both National and Higher National awards such as National Certificate (NC) and National Diploma (ND), and Higher National Certificate (HNC) and Higher National Diploma (HND) in Furniture Making, which are very useful for those wishing to become cabinet-makers.

FT2 – FILM AND TELEVISION FREELANCE TRAINING

If you'd like to work in the glamorous world of showbiz, this could be right up your street. The FT2 scheme offers a two-year construction apprenticeship for people wishing to work on film and television sets. Obviously, the skills of a carpenter are essential for building sets and there is always plenty of work in England's busy studios such as Elstree and Pinewood. You have to be 16 or over to qualify for the scheme and will probably intersperse block release from college with actual work, eventually gaining an NVQ 3 qualification. (For more details see the Resources chapter, page 57.)

Most carpenters and cabinet-makers are trained to NVQ/SVQ levels 2 and 3, but if you want to gain promotion to a management position, such as site manager or contract manager, you can train up to NVQ/SVQ levels 4 and 5. If you wish to take your education even further you can do a Foundation Degree, which is a mix of vocational and academic learning. Generally, a Foundation Degree will take two years if you study full-time, or three years studying part-time. Once you have a job in the industry, you can take a number of short courses that cover specific areas, such as the one-day CITB-approved Wood Machines Safety Course designed for machine operators. Your employer will usually pay for this type of course.

For most of the courses above there are no set entry qualifications, although it is good to be able to show a certain level of academic achievement. Good GCSEs to have are Maths, English, Science, and a craft or practical subject such as Woodwork or Needlework.

However, there are set entry qualifications for both the SQA courses and for the BTEC courses. For the SQA awards you will need four GCSE passes, grades A to C, with at least one A pass. For BTEC you will need four GCSE passes, grades A to C.

(BTEC is administered by Edexcel, which is currently being reorganised, and all entry qualifications should be checked with them.)

Getting a good academic record isn't the only positive thing you can do while at school to help your entry into the profession. One of the best things you can do is to get some work experience. As so many companies are eager to get the best trainees, they now offer work experience as a 'taster', letting you see what they actually do, and maybe encouraging you to do your apprenticeship with them. Talk to your careers officer or careers advisor about work experience opportunities in your area. Also ask them if your school is due to be visited by CITB. In 2002 CITB undertook activities with over 6,000 schools, and each October CITB organises National Construction Week, which includes classroom visits by people who actually work in the industry.

> **DID YOU KNOW?**
>
> Toyah Willcox's dad once owned a successful joinery business in Birmingham.

If you have the facilities, actually building something is an excellent way to show off your skills. It need not be anything particularly complicated or big, but if it is well made and well finished it will show you have a talent for carpentry. If you are doing woodwork at school, this will be easy for you. If this is not practical, you can always show your interest in the subject by keeping abreast of new ideas, design trends and industry news by reading one of the many magazines dedicated to carpentry and cabinet-making. Details of the major publications are listed in the Resources chapter, page 57.

Whichever way you decide to train, finding the best training college for your chosen career is crucial. Many colleges have now become Centres of Vocational Excellence, which means they specialise in particular subjects. For instance, Lambeth

College is the COVE for the construction industry in London. Ask your careers officer if there is one in your area. You should also ask about the National Construction College. This is a network of colleges around the UK dedicated to training skilled construction workers (see the Resources chapter). Also find out which courses are actually available in your area. For example, at the Building Craft College in Stratford they offer: an MA in Shopfitting; City and Guilds Diploma in Fine Woodwork; NVQ in Wood Occupations; and a one-day, CITB-approved course in Wood Machines Safety. Other courses include Veneering, Bench Joinery and Machine Woodworking. Check to see what would be most suitable for you by looking at the websites of the main awarders of vocational qualifications (City and Guilds, CITB, and Edexcel in England, Scottish Qualifications Authority (SQA) in Scotland). All these addresses are included in the Resources chapter.

The guide on page 46 neatly sums up the various routes into a career in carpentry and cabinet-making, from the time you leave school, right up to the highest levels.

Well trained

What you learn as a trainee will very much depend on what branch of carpentry you decide to go in to. For instance, a cabinet-maker won't be up on a roof fitting beams, while a shopfitter won't need to know how to do marquetry. However, there are some aspects of working with wood that all trainees need to know about. These include:

- **Product recognition**

One of the most basic skills anyone working with wood needs is how to distinguish between different woods; their particular properties; what they are used for; and how to preserve them. For instance, hardwoods can be very expensive and so are used only for certain projects. Other woods, such as walnut, are used for their decorative qualities. At college you will be taught how to recognise a great variety of woods and how they can be preserved.

- **Use of tools**

Knowing which tools to use for which job is an essential skill. In a workshop there is a huge range of tools, from saws and screwdrivers to different hammers, chisels and planers. You will be taught how to use them effectively and, above all, safely.

- **Health and safety**

Health and safety is of the utmost importance in carpentry. Not only do you need to know how to use your tools safely, but also to be aware of safety issues in your environment including wood dust, certain chemicals such as adhesives, paints and protective chemicals, and structural security. You will be taught the proper use of safety equipment such as facemasks and earplugs.

- **Fixings and hardware**
Carpentry isn't just about wood, it is also about the fixings, such as screws and nails, used to hold items together or attach them to other things, and it is also about hinges, plates, brackets, locks, and handles. You will need to know what each different piece of hardware does and which pieces you need for each different job.

- **Basic craft skills**
As a trainee you will be required to do a number of set-piece assessment projects where you actually make things. These will test your ability to do geometry, measure correctly, and put a wooden structure together properly. At NVQ Level 2 projects usually include putting together a full-size door and door frame, as well as making a cabinet of some description. As your training advances, so do the projects, to a level where you may have to complete a box sash window or a model staircase. Projects will vary depending on which course you are taking. For instance a City and Guilds Diploma in Fine Woodwork includes practical experience with woodturning, hand veneering, marquetry and French polishing, while students doing an MA in shopfitting must have practical knowledge of machining from sawn stock, making a curved bench seating unit, and finishing complex internal panelling.

- **Business skills**
Once again, depending on which course you are taking, you may well have to learn business skills such as estimating quantities, producing cutting lists, and measuring up and producing scale drawings based on a client's brief. Your course may also include key skills in numeracy, literacy, and information technology, all of which will be essential to you, especially if you decide to go freelance and set up your own business.

access to
CARPENTRY AND CABINET-MAKING

NO QUALIFICATIONS

ENTRY LEVEL QUALIFICATION

FOUR GCSEs (A-D) grades 1-3
GNVQ/GSNVQ level 1
selection interview

ON THE JOB TRAINING

APPRENTICESHIP ✦ TRAINEE SCHEMES

ADVANCED MODERN APPRENTICESHIP (England)
SKILLSEEKERS (Scotland)
MODERN APPRENTICESHIP (NI)
MODERN APPRENTICESHIP (Wales)

e.g. CITB (Construction Industry Training Board)
NAS (National Association of Shopfitters)

e.g.
JOINER
FORMWORKER
CARPENTER
SHOPFITTER

CREDITS/FURTHER LEARNING

ON THE JOB QUALIFICATIONS ✦ PROFESSIONAL BODIES

NVQ/SNVQ level 1
BTEC HNC/HND
Full-time/part-time/distance learning

e.g. GUILD OF MASTER CRAFTSMEN

CAREER OPPORTUNITIES

DEVELOPMENT OPTIONS

HIGHER EDUCATION ✦ MANAGEMENT ✦ FREELANCE

- **Fixings and hardware**
Carpentry isn't just about wood, it is also about the fixings, such as screws and nails, used to hold items together or attach them to other things, and it is also about hinges, plates, brackets, locks, and handles. You will need to know what each different piece of hardware does and which pieces you need for each different job.

- **Basic craft skills**
As a trainee you will be required to do a number of set-piece assessment projects where you actually make things. These will test your ability to do geometry, measure correctly, and put a wooden structure together properly. At NVQ Level 2 projects usually include putting together a full-size door and door frame, as well as making a cabinet of some description. As your training advances, so do the projects, to a level where you may have to complete a box sash window or a model staircase. Projects will vary depending on which course you are taking. For instance a City and Guilds Diploma in Fine Woodwork includes practical experience with woodturning, hand veneering, marquetry and French polishing, while students doing an MA in shopfitting must have practical knowledge of machining from sawn stock, making a curved bench seating unit, and finishing complex internal panelling.

- **Business skills**
Once again, depending on which course you are taking, you may well have to learn business skills such as estimating quantities, producing cutting lists, and measuring up and producing scale drawings based on a client's brief. Your course may also include key skills in numeracy, literacy, and information technology, all of which will be essential to you, especially if you decide to go freelance and set up your own business.

access to
CARPENTRY AND CABINET-MAKING

NO QUALIFICATIONS

ENTRY LEVEL QUALIFICATION

FOUR GCSEs (A-D) grades 1-3
GNVQ/GSNVQ level 1
selection interview

ON THE JOB TRAINING

APPRENTICESHIP ♦ TRAINEE SCHEMES

ADVANCED MODERN APPRENTICESHIP (England)
SKILLSEEKERS (Scotland)
MODERN APPRENTICESHIP (NI)
MODERN APPRENTICESHIP (Wales)

e.g. CITB (Construction Industry Training Board)
NAS (National Association of Shopfitters)

e.g.
JOINER
FORMWORKER
CARPENTER
SHOPFITTER

CREDITS/FURTHER LEARNING

ON THE JOB QUALIFICATIONS ♦ PROFESSIONAL BODIES

NVQ/SNVQ level 1
BTEC HNC/HND
Full-time/part-time/distance learning

e.g. GUILD OF MASTER CRAFTSMEN

CAREER OPPORTUNITIES

DEVELOPMENT OPTIONS

HIGHER EDUCATION ♦ MANAGEMENT ♦ FREELANCE

PAUL MAGERUM

Case study 2

8

THE MAINTENANCE CARPENTER
At 43 years old, Paul Magerum is the Maintenance Carpenter for Writtle Agricultural College in Chelmsford. As a boy he was interested in woodwork and he wasn't very academic, so he decided to go into the building trade. At 15 he started a three-year apprenticeship with a bespoke joiners based in Stratford. He worked on the job and went to college one day and two evenings a week. He started off doing menial jobs for the firm, but at the end of his apprenticeship he was doing everything the other tradesmen were doing. He particularly liked going out from the workshop to fix whatever had been made on site in places such as University College London, office blocks and in private homes.

Paul stayed with the firm for 10 years before leaving to become self-employed. He mainly worked on refurbishing buildings and became involved with a firm that started using him on a fairly regular basis. One of the firm's clients was the merchant bank Kleinwort Benson and he worked on their buildings as a maintenance carpenter for the next seven years. He particularly enjoyed working on the older buildings where he had to fix and replace wood panelling and oak doors. For the last two years at Benson he

> I like the creating and the satisfaction at the end of the day when you can see what you have built.

was the foreman. He then worked for another banking firm, Merrill Lynch, initially with responsibility for one building, but after seven years he was leading a large team and looking after 13 buildings.

Realising this was not what he wanted to do, he left to become self-employed again before getting the position at the agricultural college, which is on a 500-acre farm. Paul is responsible for the upkeep of 80 buildings, from sheds to office blocks. Last year, he built a wooden bridge from scratch and now everyone at the college uses it daily.

> There are so many different aspects to carpentry you really need to be a good all-rounder to make a go of it.

'When I left my first employers to go self-employed I suddenly started earning about three times the amount of money I earned before and it is good to bear in mind that workshop-based carpenters and joiners are probably paid the least in the industry. Initially, I enjoyed working with a big team at Kleinwort Benson, but towards the end of my time with Merrill Lynch I was off the tools and basically doing a desk job and I slowly began to realise it wasn't me. It was slowly killing me and so I reassessed my life. I realised I missed the hands-on stuff and I wasn't enjoying what I was doing. Now I'm back using tools again, using my hands, and as there are so many buildings here the work is very varied. One day I'll be putting up a partition wall, or working on a barn, or even building a roof.

'I like the creating and the satisfaction at the end of the day when you can see what you have built. I really like working with my hands and with tools and the good thing the Merrill

Lynch job taught me was that I am not a boss or a manager. I like working by myself on projects, I prefer it so that is one aspect of this job that really suits me. Plus, it is only five minutes from where I live so I don't have to commute like I did when I worked in the City, and also, I am out in the fresh air all day.

'Communication is definitely one of the skills you need for this job. Because of my experience I can communicate with everyone from the labourer right up to the managing director. You also have to really enjoy what you are doing. There are so many different aspects to carpentry you really need to be a good all-rounder to make a go of it. There is a massive, massive shortage and you can make a really good living from it, but you've just got to really want to do it. For me it is a passion, like being a chef, and I'm enjoying what I am doing now more than I ever have and that's saying a lot considering I started at 15.

'It was a fantastic feeling when I finished that bridge and now, a year later, I've just given it a coat of linseed oil and everyone who walks over it comments on it. How many people get a chance to do stuff like that? Knowing what I know now, if I was starting out in carpentry again I would concentrate on working with traditional tools on traditional buildings because that is what I am interested in. For instance, if I had an opportunity to work on the reconstructed Shakespeare's Globe theatre, well that would just be a dream come true for me. There are those fantastic jobs out there, and that's what I'd do.'

Career opportunities

Earlier, in the What's The Story? chapter, we looked at just some of the jobs open to trained carpenters. Even before you start to train it is a good idea to look at what job opportunities could be there for you in the future. The skills you learn as you train will stay with you for the rest of your life and they can take you right to the top of your profession. Take the career path of Richard Easton, for example. He started off at 16 as an Apprentice Joiner doing his City and Guilds at A. E. Hadley Ltd. He was then promoted to Setter Out, became an Estimator with the firm and rose to Director of Sales and Overseeing Estimating. After a spell as a Co-Director he went on to become Managing Director of the whole firm. Your ambitions may not be this high, but there will be plenty of different career paths for you to take. The following diagram will give you a rough idea of just how far you can go.

We have already discussed what formworkers, joiners, shopfitters, and cabinet-makers actually do but as you advance up the career ladder, your role and your responsibilities will obviously change. For a start, you may find your job becomes less 'hands on' and more involved with paperwork and co-ordination. For example, **setters out** are very skilled shopfitters who prepare working, full-size or scale drawings of the fittings to be made, while the actual cutting and fitting together of the cabinets or shelving units is carried out by someone else. **Estimators** work out the amount of raw materials that will be needed for each job and price up the contracts. This is extremely important work, for if the estimator gets it wrong, he or she could lose contracts for their company.

This is an industry with real opportunities for advancement. As you go through the training and discover where your strengths lie you will be able to map out a future career path. The diagram below shows options that will open up to you once you have trained.

CAREER OPPORTUNITIES

TRAINING IN WOODWORKING SKILLS

FORMWORKER ♦ CARPENTER ♦ JOINER/BENCH JOINER ♦ SHOPFITTER

MORE EXPERIENCE
NVQ LEVEL 3 UPWARDS

MAINTENANCE CARPENTER ♦ SELF-EMPLOYED CARPENTER
BUILDING CONSERVATION & RESTORATION ♦ BESPOKE CABINET-MAKER
SETTER OUTER

FURTHER EXPERIENCE
NVQ LEVEL 3/4 TRAINING FOR MANAGEMENT

SUPERVISOR/TEAM LEADER ♦ ESTIMATOR

FURTHER EXPERIENCE
NVQ LEVEL 4 & 5

SITE MANAGER ♦ WORKING FOREMAN/MANAGER ♦ CONTRACT MANAGER

When you get to management level you may find you do no actual woodwork at all, but spend your days on the phone to suppliers, co-ordinating jobs with other people working on your site such as plumbers and electricians, communicating with your staff, and keeping your clients informed of progress. You will also have to sort out any problems that may arise, such as a job going seriously over-budget. **Supervisors**, **site managers** and **team leaders** need exceptionally good people skills as they oversee other people's work on site and have to offer advice, make suggestions and sort things out when a job goes wrong. They must also be very aware of the health and safety aspects of the job, for example ensuring workers wear safety gear such as steel-toed boots and helmets while on site. **Working foremen** or **managers** are responsible for running workshops and looking after the joiners who work there. Quality control is a big part of their job and so they have to pay attention to detail. They also have to ensure the finished product is made and delivered on time: many clients insert time-sensitive clauses in their contracts, which means that if their order is not delivered on time they pay less. Once again, they must be very conscious of health and safety, as joiners use a lot of power tools. **Contract managers** have many different responsibilities. They have a supervisory role on site and oversee the whole project, from the planning stage right through to the final product. They have to make sure that even the smallest details in the contract are not overlooked and will liaise closely with the client.

As you progress up the promotion ladder, you will dramatically increase your earning potential, so it makes sense to get the best training you can. Remember, a carpenter's skills are highly regarded and their services much sought after, so if this is the career you have chosen to pursue make sure you take advantage of all the opportunities to learn that are available to you.

The last word

10

By now you should have a pretty good idea what working as a carpenter or a cabinet-maker actually entails. Mankind has been working with wood since the dawn of civilisation and even with the advent of man-made materials such as plastic and concrete the skills of the woodworker are still very much in demand. Deciding on a career in carpentry will open up a world of opportunities for you. Overleaf is a fun checklist to see if carpentry really is the career choice for you.

There has never been a better time to train to become a carpenter. With a shortage of skilled craftsmen across the whole industry, well-trained, professional carpenters are now at a premium. This is varied, interesting and above all fulfilling work where you can actually see the fruits of your labours. It can also be very rewarding financially, with very talented craftsmen earning substantial amounts for their expertise. There are also plenty of opportunities to advance to management level and beyond. The future for the whole industry is looking bright and hopefully this book has helped you to make up your mind whether or not you want to be part of that industry. The following chapter contains comprehensive information on the organisations you should contact if you want to make a career in carpentry.

If you have made it this far through the book then you should know if **Carpentry and Cabinet-Making** really is the career for you. But, before contacting the professional bodies listed in the next chapter, here's a final, fun checklist to show if you have chosen wisely.

THE LAST WORD ✔ TICK YES OR NO

DO YOU LIKE WORKING WITH YOUR HANDS?
☐ YES
☐ NO

DO YOU CONSIDER YOURSELF CREATIVE?
☐ YES
☐ NO

DO YOU WANT A JOB WHERE YOU WILL BE DOING SOMETHING DIFFERENT EVERYDAY?
☐ YES
☐ NO

ARE YOU SELF MOTIVATED AND ABLE TO THINK ON YOUR FEET?
☐ YES
☐ NO

ARE YOU ABLE TO COMMUNICATE EFFECTIVELY WITH PEOPLE?
☐ YES
☐ NO

ARE YOU A SELF STARTER, ABLE TO TAKE CONTROL AND RESPONSIBILITY?
☐ YES
☐ NO

If you answered 'YES' to all these questions then
CONGRATULATIONS! YOU'VE CHOSEN THE RIGHT CAREER!
If you answered 'NO' to any of these questions then this may not be the career for you.
However, there are still some options open to you,
for example, you could work as a Setter Outer or a DIY Shop Assistant

Resources

11

In this section you will find all the addresses, telephone numbers and websites for the relevant government and industry advisory and training bodies for the construction industry. There is also a list of publications you may find useful to read.

TRAINING AND ADVICE

City and Guilds
1 Giltspur Street
London EC1A 9DD
020 7294 2468
www.city-and-guilds.co.uk

City and Guilds is the leading provider of vocational qualifications in the United Kingdom. It has five different levels of qualification, with 1 based on the lowest competence level, and 5 based on the highest, and it offers everything from NVQ and SVQ to Modern Apprenticeships and Higher Level Qualifications. The excellent website lists all the qualifications it provides in carpentry and cabinet-making (search under Construction).

Connexions
www.connexions.org.com
www.connexionscard.com

The Connexions service has been set up especially for 13- to 19-year-olds and offers advice, support and practical help on many subjects including your future career options. In the Career Zone on the site you will find Career Bank, offering information on training and jobs in carpentry and cabinet-making.

Construction Industry Training Board (CITB)
Bircham Newton
King's Lynn

Norfolk PE31 6RH
01485 577577
www.citb.co.uk

The CITB runs most of the employer-based training in the UK and is committed to training professionals to a very high level. If you wish to get a place on its Construction Apprenticeship Scheme you can apply to the CITB direct. It has some very good pamphlets and brochures offering more information, including a sheet entitled Wood Occupations and brochures on Shopfitting and on Building Conservation and Restoration. Alternatively have a look at the website.

In Scotland:
4 Edison Street
Hillington
Glasgow G52 4XN
0141 810 3044

In Wales:
Units 4 and 5, Bridgend Business Centre
David Street
Bridgend Industrial Estate
Bridgend CF31 3SH
01656 655226

Department for Education and Skills (DfES)

Packs available from 0800 585505
www.dfes.gov.uk

If you are undertaking a vocational training course lasting up to two years (with one year's practical work experience if it is part of the course) you may be eligible for a Career Development Loan. These are available for full-time, part-time and distance learning courses and applicants can be employed, self-employed, or unemployed. The DfES pays interest on the loan for the length of the course and up to one month afterwards. You can also find out which colleges are Centres of Vocational Excellence at the website www.dfes.gov.uk.coves.

Edexcel
Stuart House
32 Russell Square
London WC1B 5DN
0870 240 9800
www.edexcel.org.uk

Edexcel has taken over from BTEC in offering BTEC qualifications including BTEC First Diplomas, BTEC National Diplomas, and BTEC Higher Nationals (HNC and HND). It also offers NVQ qualifications. The website includes qualification 'quick links' and you can search by the qualification or the career you are interested in. Edexcel is currently being reorganised and all course and qualification information should be checked with them.

Federation of Master Builders
14–15 Great James Street
London WC1N 3DP
020 7242 7583
www.fmb.org.uk

This is the largest trade association for the UK construction industry and represents over 13,000 small and medium-sized building companies.

FT2 – Film and Television Freelance Training
www.ft2.org.uk

Offers two-year construction apprenticeships for people aged 16 and over who want to work on film and TV sets.

Guild of Master Craftsmen
166 High Street
Lewes
East Sussex BN7 1XU
01273 478449
www.guildmc.com

Thousands of companies up and down the country belong to the Guild, which endeavours to maintain and uphold standards of excellence within the industry.

Institute of Carpenters (central office)
35 Hayworth Road
Sandiacre
Nottingham NG10 5LL
0115 949 0641
www.carpenters-institute.org

The IOC is a craft association for everybody working in the wood trades.

Learning and Skills Council
Modern Apprenticeship helpline
08000 150600
www.lsc.gov.uk
www.realworkrealpay.co.uk

Launched in 2001, Learning And Skills Council now has 48 branches across the country. It is responsible for the largest investment in post-16 education and training in England and this includes further education colleges, work-based training and workforce developments. Its realworkrealpay website is specifically aimed at those who would like to do Modern Apprenticeships.

For MAs in Scotland you should look at

www.modernapprenticeships.com or
www.careers.scotland.org.uk.

In Wales you should look at www.beskilled.net.

National Association of Shopfitters (NAS)
NAS House
411 Limpsfield Road
The Green
Warlingham
Surrey CR6 9HA
01883 624961
www.shopfitters.org

NAS welcomes applications for training from all eligible young people, regardless of sex, race or disability. It runs its own annual awards competitions, including the Second Year Apprentice Award.

National Construction College (NCC)
01485 577669
E-mail: direct.training@citb.co.uk

The NCC is a network of colleges around the country that specialise in training young people as skilled operatives and potential supervisors in the construction industry. All courses combine a mixture of site work and residential training and last between 4 and 43 weeks.

New Deal
www.newdeal.co.uk

If you are an older individual looking to change careers and you have been unemployed for six months or more (or receiving Jobseekers Allowance), you may be able to gain access to NVQ/SVQ courses through the New Deal Programme. People with disabilities, ex-offenders and lone parents are eligible before reaching six months of unemployment. Check out the website for more information.

Qualifications and Curriculum Authority (QCA)
83 Piccadilly
London W1J 8QA
020 7509 5555
www.qca.org.uk

In Scotland:
Scottish Qualifications Authority (SQA)
Hanover House
24 Douglas Street
Glasgow G2 7NQ
Customer Contact Centre: 0141 242 2214
www.sqa.org.uk

These official awarding bodies will be able to tell you whether the course you choose leads to a nationally approved qualification such as NVQ or SVQ.

Worshipful Company of Carpenters (The Carpenters' Company)
Carpenters' Hall
Throgmorton Avenue
London EC2N 2JJ
020 7588 7001
www.the carpenterscompany.co.uk

One of the City of London's oldest livery companies, The Carpenter's Company was founded in 1271. It runs the Building Crafts College and provides scholarships in the trade as well as sponsoring craft competitions.

Building Crafts College
Kennard Road

Stratford
London E15 1AH
020 8522 1705

PERIODICALS

Cabinet Maker
7th floor
Ludgate House
245 Blackfriars Road
London SE1 9UR
0207 921 8406
www.cm1st.com

This weekly publication is one of the best-known titles for the industry.

Furniture and Cabinet Making
86 High Street
Lewes BN7 1XN
01273 477374
www.gmcmags.com

The monthly magazine of the Guild of Master Craftsmen: aimed at makers of bespoke fine furniture.

Woodworking News
Old Sun
Crete Hall Road
Northfleet DA11 9AA
01474 536536
www.nelton.co.uk

This magazine is published ten times a year and is aimed at craftsmen, giving product and industry news. Nelton, the publishing company, has two other titles that may be of interest: *Irish Woodworking and Furniture News*; and *Furniture Products*.